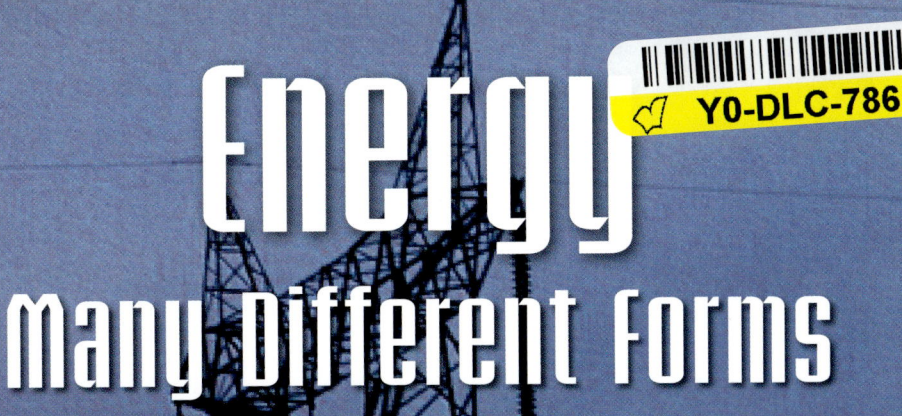

Energy
Many Different Forms

by Lisa A. Boehm

Table of Contents

Develop Language 2

CHAPTER 1 Defining Energy 4
　　　　　　　Your Turn: Communicate 9
CHAPTER 2 Different Forms of Energy 10
　　　　　　　Your Turn: Classify 15
CHAPTER 3 More Forms of Energy 16
　　　　　　　Your Turn: Interpret Data 19

Career Explorations 20
Use Language to Explain 21
Science Around You 22
Key Words 23
Index 24

DEVELOP LANGUAGE

Lots of things happen in a big city. Lights and sounds are everywhere. Everything is in **motion**. All of these things need **energy**.

Look at the pictures. Talk about what you see.

Why does the taxi need energy?

The taxi needs energy to _____.

What else in the photos uses energy?

The _____ uses energy to _____.

Name other ways that a city uses energy.

motion – the action of moving or being moved

energy – the ability to do work or cause change

chef

2 Energy: Many Different Forms

Develop Language 3

CHAPTER 1

Defining Energy

Energy is a very important part of our world. It takes energy to make anything happen.

Your body uses energy to run, jump, and play games. You use energy to bounce and throw a ball.

Whenever energy is used, a change occurs. Energy is the ability to cause change.

◀ Your body uses energy when you play basketball.

Energy: Many Different Forms

A student uses **force** to pick up a backpack. A force is either a push or a pull. If she pulls upward with enough force, the backpack will move upward. When a force moves an object in the **direction** of the force, **work** is done.

It takes energy to do the work of lifting the backpack. So, we can say that energy is the ability to do work or cause change.

force – a push or a pull
direction – the way in which an object is moving
work – the result of a force moving an object

▼ You need more energy to lift a backpack full of books.

▲ You need less energy to lift an MP3 player.

KEY IDEA Energy is the ability to do work or cause change.

Chapter 1: Defining Energy 5

Kinetic Energy

There are two kinds of energy. One kind is **kinetic energy**. All moving objects have kinetic energy. Kinetic energy is the energy of motion.

Different moving objects have different amounts of kinetic energy. Larger, faster objects have more kinetic energy than smaller, slower objects.

kinetic energy – the energy of motion

A leaping dolphin has a large amount of kinetic energy.

A crawling crab has a small amount of kinetic energy.

6 *Energy: Many Different Forms*

Potential Energy

The other kind of energy is **potential energy**. Potential energy is stored energy, or energy that is not being used.

An object can have potential energy because of its **position**. A vase on a table can tip over or fall to the floor. The vase has potential energy because of its position.

An object can also have potential energy because of its **condition**. The bent bow has potential energy because it will straighten when the boy lets the arrow fly. The bow has potential energy because of its condition.

▼ Pulling the string changes the bow's shape. The bent bow stores potential energy.

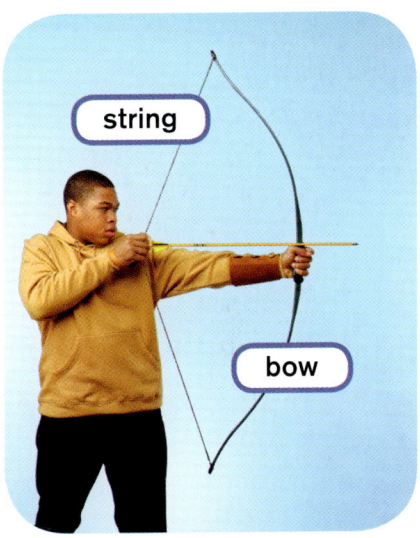

potential energy – stored energy
position – where an object is located
condition – an object's shape or what it contains

▼ The vase can fall. It has potential energy because of its position.

KEY IDEA Energy can be either kinetic energy or potential energy.

Chapter 1: Defining Energy 7

Energy Changes

An object's energy can change from potential energy to kinetic energy, or from kinetic energy to potential energy.

The cyclist has potential energy when he is at the top of the hill. As he moves downward, his potential energy is changing to kinetic energy.

When he moves up the next hill, his kinetic energy will change back into potential energy.

Explore Language

Your **potential** is what you will be able to achieve in the future.

If an idea has **potential**, it has promise and is worth exploring further.

SHARE IDEAS Explain how a ball could have potential energy. Explain how this energy could change to kinetic energy.

KEY IDEA An object's energy can change from kinetic energy to potential energy, or from potential energy to kinetic energy.

The cyclist's potential energy is changing to kinetic energy.

Energy: Many Different Forms

YOUR TURN

COMMUNICATE

Draw a diagram of this roller coaster.

1. Where will the roller coaster cars have the most kinetic energy? Mark and label that position on your diagram.

2. Where will the roller coaster cars have the most potential energy? Mark and label that position on your diagram.

Discuss your diagram with a friend.

MAKE CONNECTIONS

The balloons have potential energy because they are stretched and full of air. What will happen to the balloons' potential energy if they are popped with a pin? How do you know?

USE THE LANGUAGE OF SCIENCE

What are two ways that an object can have potential energy?

An object can have potential energy because of its position or its condition.

Chapter 1: Defining Energy

CHAPTER 2

Different Forms of Energy

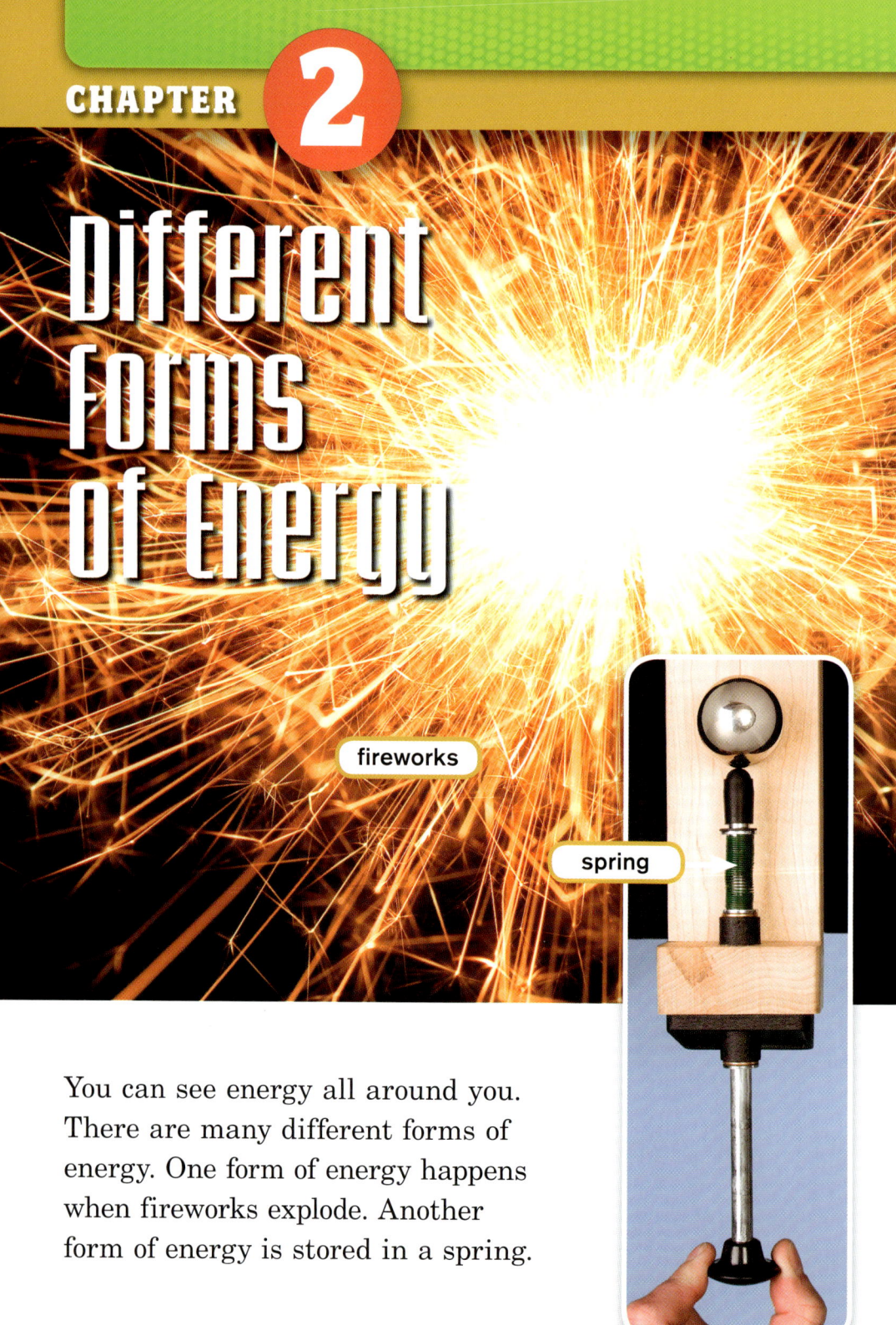

fireworks

spring

You can see energy all around you. There are many different forms of energy. One form of energy happens when fireworks explode. Another form of energy is stored in a spring.

Mechanical Energy

Mechanical energy is the energy that an object has to do work. Mechanical energy is made up of two parts. One part comes from the object's kinetic energy. The other part comes from the object's potential energy. The mechanical energy of an object is the sum of its kinetic energy and its potential energy.

mechanical energy – the sum of an object's kinetic energy and potential energy

▼ Part of the breaking wave's mechanical energy comes from its kinetic energy. The rest of the breaking wave's mechanical energy comes from its potential energy.

Chapter 2: Different Forms of Energy 11

Sound Energy

Sound energy is a form of energy made by **vibrations**, quick back-and-forth movements of the particles that make up **matter**. Sound energy does work when it makes matter move.

If you clap your hands, you set up vibrations in the air around your hands. These vibrations travel to your ears, and you hear sound.

sound energy – a form of energy made by vibrations
vibrations – quick back-and-forth movements
matter – anything that has mass and takes up space

By The Way...

Sometimes you may feel vibrations from a loud sound. For example, you can feel the ground vibrate when a plane passes overhead.

Chemical Energy

Chemical energy is a form of energy stored inside the particles that make up **substances**. When the particles break apart, the chemical energy is released, or freed.

Chemical energy is stored in fuels like wood, coal, and oil. Chemical energy is also stored in batteries, fireworks, and even the food you eat.

chemical energy – a form of energy stored inside the particles that make up matter

substances – kinds of matter with a certain set of properties

When rocket fuel burns, stored energy is released and the rocket blasts off.

Chapter 2: Different Forms of Energy

Thermal Energy

The tiny particles in matter are always moving. That means they always have energy. The energy of the particles is **thermal energy**.

The amount of thermal energy in a substance is different under different conditions. For example, the particles in a cup of hot water move quickly. The particles in a cup of cold water move less quickly. So a cup of cold water has less thermal energy than a cup of hot water.

> **thermal energy** – a form of energy that is the result of the movement of particles of matter

The particles in hot water move quickly.

The particles in cold water move less quickly.

> **KEY IDEA** Mechanical energy, sound energy, chemical energy, and thermal energy are all forms of energy.

14 *Energy: Many Different Forms*

YOUR TURN

CLASSIFY

Classify the photos. Make a chart like this one. For each photo, mark the forms of energy it shows. Each photo may show more than one form of energy. Discuss your chart with a friend.

Form of Energy	sugar burning	water falling	whistle blowing
mechanical			
sound			
chemical			
thermal			

MAKE CONNECTIONS

Athletes are careful about the food they eat before a race or game. They want to eat food that will give them the greatest amount of energy. Think about a bowl of steaming rice. What forms of energy does the rice have?

 STRATEGY FOCUS

Determine Importance

Look at this chapter again. What are the most important ideas? How do you know?

Chapter 2: Different Forms of Energy

CHAPTER 3

More Forms of Energy

Most of the energy we use on Earth comes from the sun. The sun's energy is the result of changes inside the particles that make up the sun.

Plants and animals depend on energy from the sun to live and grow.

tropical plant

green iguana

Light energy from a laser does work as it cuts through metal.

Light Energy

Light is energy we can see. **Light energy** is a form of energy stored inside the particles of some kinds of matter. Changes inside the particles release light energy.

Some light sources are natural, such as sunlight and fireflies. Other light sources are **artificial**, or made by people. Artificial light sources include lamps, flashlights, and lasers.

light energy – a form of energy stored inside the particles of some kinds of matter

artificial – made by people

Electrical Energy

Electrical energy is another form of energy stored inside the particles of some kinds of matter.

Electrical energy gives us an easy way to move energy from one place to another. We use electrical energy to watch television or to use a computer. Most of this energy is produced in **power plants**. The power plants send the electrical energy through power lines to our homes. Power plants use many different materials to produce electrical energy.

electrical energy – a form of energy stored inside the particles of some kinds of matter

power plants – places where electric energy is produced

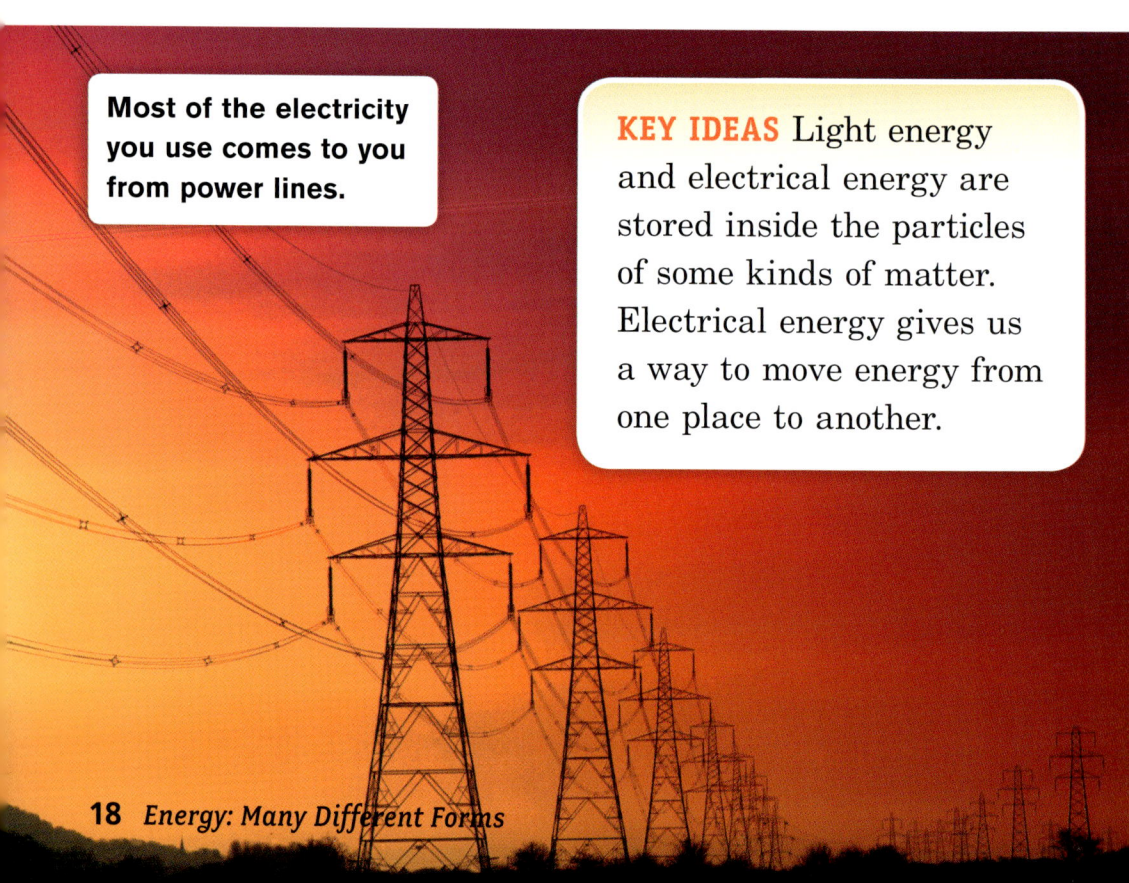

Most of the electricity you use comes to you from power lines.

KEY IDEAS Light energy and electrical energy are stored inside the particles of some kinds of matter. Electrical energy gives us a way to move energy from one place to another.

18 *Energy: Many Different Forms*

YOUR TURN

INTERPRET DATA

Power plants in the United States use different fuels to make electrical energy. Look at the chart. Interpret the data to answer the questions.

1. What fuel does the United States use the most to make electrical energy?
2. How much of electrical energy used in the United States comes from natural gas?

Materials Used to Produce Electrical Energy in the United States (2005)	
coal	51%
natural gas	17%
moving water (hydropower)	7%
petroleum (oil)	3%
other	22%

MAKE CONNECTIONS

Make a list of all the ways you use electrical energy in a day. Share your list.

EXPAND VOCABULARY

The word **power** is used in science, social studies, and math. It can mean different things. Use the word **power** in different ways by answering these questions.

- Why does a city need electrical **power**?
- Why is it important to have political **power**?
- How much is 3 to the second **power**?

Chapter 3: More Forms of Energy

CAREER EXPLORATIONS

What Is an Electrician?

Electricians bring electrical energy into homes and businesses. They also fix electrical machines.

- Read the chart.
- Would you like to be an electrician? Explain your answer.

Would I like this career?	You might like this career if: • you are interested in how electrical energy works. • you like to work with your hands and fix things. • you like to study diagrams and blueprints.
What would I do?	• You would install electrical wiring in homes and businesses. • You would fix electrical machines or broken wiring.
How can I prepare for this career?	• Most electricians begin as apprentices. Apprentices are paid while they learn from experienced electricians. • Apprentices usually need a high school diploma. They take classes to learn about electronics and electrical energy. • Apprentices can work on their own after 3 to 5 years.

20 *Energy: Many Different Forms*

USE LANGUAGE TO EXPLAIN

Words that Explain

When you explain, you make something easier to understand. One way to explain is to tell **why** something happens. Words such as **because**, **so**, and **since** help tell why something happens.

EXAMPLES

A person has potential energy at the top of a hill **because** he has the potential to move down the hill.

The balloon is stretched and full of air, **so** it has potential energy.

Since the river is moving, it has kinetic energy.

Look at the photographs on page 15. Explain why each photograph is different. Use **because**, **so**, or **since**.

Write to Explain

Choose a photo in this book or another photo. Write about the form of energy that is shown in the photo.

- Describe the photo. Tell what is moving.
- Explain why the object is moving.
- Tell what form of energy is being used.

Words You Can Use	
because	since
so	why

SCIENCE AROUND YOU

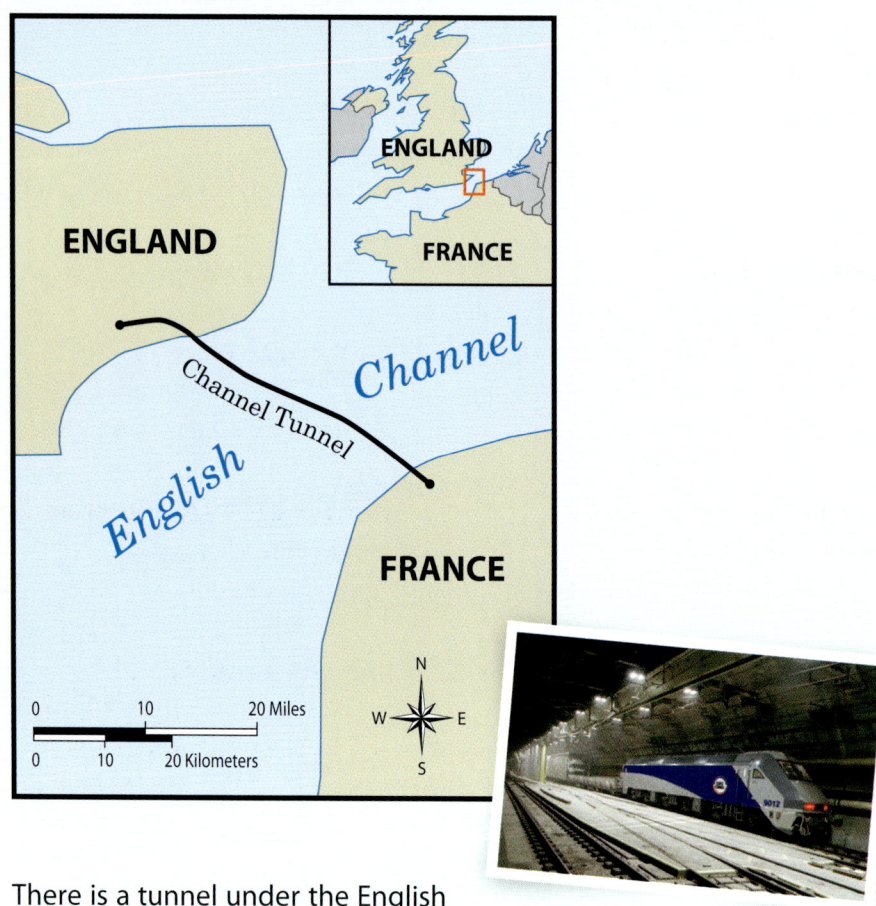

There is a tunnel under the English Channel. The tunnel connects England and France. Cars and trucks can travel through the tunnel. The Eurotunnel Freight shuttle also uses the tunnel.

Look at the map. Answer these questions.

- About how long is the English Channel Tunnel?

- If the train travels at 300 kilometers per hour (186 miles per hour), about how long will it take to cross the channel?

- What has more kinetic energy, a car or the train? Explain your answer.

22 *Energy: Many Different Forms*

Key Words

chemical energy a form of energy stored inside the particles that make up matter
Burning wood releases the **chemical energy** in the wood.

electrical energy a form of energy stored inside the particles of some kinds of matter
Televisions need **electrical energy** to work.

energy the ability to do work or cause change
The sun is Earth's most important source of **energy**.

force (forces) a push or a pull
You use **force** when you pick up a book.

kinetic energy the energy of motion
All moving objects have **kinetic energy**.

light energy a form of energy stored inside the particles of some kinds of matter
Sometimes the particles of some kinds of matter release **light energy.**

mechanical energy the sum of an object's kinetic energy and potential energy
The **mechanical energy** of an object includes its kinetic energy and its potential energy.

motion the action of moving or being moved
Watch the **motion** of the roller coaster.

potential energy stored energy
A fully stretched rubber band has **potential energy**.

sound energy a form of energy made by vibrations
We can hear **sound energy**.

thermal energy a form of energy that is the result of the movement of particles of matter
Hot water has more **thermal energy** than ice.

work the result of a force moving an object in the direction of the force
Work is done when you drag a book across a desk.

Index

change 2, 4–5, 7–8, 16–17

chemical energy 13, 14, 15

condition 7, 9

electrical energy 18, 19, 20

direction 5

energy 2, 4–8, 10–14, 15, 16–18, 21

force 5

fuel 13, 19

kinetic energy 6, 8, 9, 11, 21, 22

light energy 17, 18

mass 12

matter 12–14, 17–18

mechanical energy 11, 14, 15

motion 2, 6

position 7, 9

potential energy 7–8, 9, 11, 21

power plant 18, 19

sound energy 12, 14, 15

stored 7, 13, 17, 18

thermal energy 14, 15

vibration 12

work 2, 5, 11, 12, 17, 20

MILLMARK EDUCATION CORPORATION
Ericka Markman, President and CEO; Karen Peratt, VP, Editorial Director; Lisa Bingen, VP, Marketing; David Willette, VP, Sales; Rachel L. Moir, VP, Operations and Production; Shelby Alinsky, Associate Editor; Ernestine Giesecke, Science Editor; Pictures Unlimited, Photo Research

PROGRAM AUTHORS
Mary Hawley; Program Author, Instructional Design
Kate Boehm Jerome; Program Author, Science

BOOK DESIGN Steve Curtis Design

CONTENT REVIEWER
Tom Nolan, Operations Engineer, NASA Jet Propulsion Laboratory, Pasadena, CA

PROGRAM ADVISORS
Scott K. Baker, PhD, Pacific Institutes for Research, Eugene, OR
Carla C. Johnson, EdD, University of Toledo, Toledo, OH
Donna Ogle, EdD, National-Louis University, Chicago, IL
Betty Ansin Smallwood, PhD, Center for Applied Linguistics, Washington, DC
Gail Thompson, PhD, Claremont Graduate University, Claremont, CA
Emma Violand-Sánchez, EdD, Arlington Public Schools, Arlington, VA (retired)

TECHNOLOGY
Arleen Nakama, Project Manager
Audio CDs: Heartworks International, Inc.
CD-ROMs: Cannery Agency

PHOTO CREDITS Cover ©Pixtal/age fotostock; IFC and 15d ©David Safanda/iStockphoto.com; 1 ©Roger Ressmeyer/CORBIS; 2-3 ©PNC/Brand X/Corbis; 2 ©Vittorio Sciosia/age fotostock; 3a ©Paul Edmondson/Corbis; 3b ©Walter Bibikow/age fotostock; 4 ©Ed Bock/CORBIS; 5a, 5b, 7a, 7b, 9c, 9d, 14a, 14b photos by Ken Karp for Millmark Education; 6a ©Stuart Westmorland/CORBIS; 6b ©Martin Harvey/CORBIS; 8 ©DiMaggio/Kalish/CORBIS; 9a ©Kord/age fotostock; 9b ©JUPITERIMAGES/Brand X/Alamy; 10a and 23 ©Suravid/Shutterstock; 10b ©Edward Kinsman/Photo Researchers, Inc.; 11 ©Kevin Schafer/age fotostock; 12 ©Dumas/agefotostock; 13 ©Ron Brown/age fotostock; 15a ©Andrew Lambert Photography/Photo Researchers, Inc.; 15b ©urosr/Shutterstock; 15c ©Symphonie/age fotostock; 16a ©photomadnz/Alamy; 16b ©Nicholas Pitt/Alamy; 17 ©Tom Tracy Photography/Alamy; 18 ©Construction Photography/Corbis; 19 ; 20 ©VStock LLC; 21 ©Sarah Johanna Eick/age fotostock; 22a map by Mapping Specialists; 22b ©Eurotunnel; 24 ©Tim Keatley/Alamy

Copyright ©2008 Millmark Education Corporation

All rights reserved. Reproduction of the whole or any part of the contents without written permission from the publisher is prohibited. Millmark Education and ConceptLinks are registered trademarks of Millmark Education Corporation.

Published by Millmark Education Corporation
PO Box 30239
Bethesda, MD 20824

ISBN-13: 978-1-4334-0196-1

Printed in the USA

10 9 8 7 6 5 4 3 2

24 *Energy: Many Different Forms*